PICTURE THIS!

In the Country

Karen Bryant-Mole

Heinemann Library

Des Plaines, Illinois

1999 Reed Educational & Professional Publishing
Published by Heinemann Library,
an imprint of Reed Educational & Professional Publishing,
1350 East Touhy Avenue, Suite 240 West
Des Plaines, IL 60018

Designed by Jean Wheeler
Commissioned photography by Zul Mukhida
Printed and bound in China

03 02 01 00 99
10 9 8 7 6 5 4 3 2 1

Library of Congress Cataloging-in-Publication Data

Bryant-Mole, Karen.
    In the country / Karen Bryant-Mole.
        p.   cm. -- (Picture this!)
    Includes index.
    Summary: Text and photographs identify various things in the country, including birds, wildflowers, insects, trees, and more.
    ISBN 1-57572-899-0 (lib bdg.)
    1. Natural history--Juvenile literature.   [1. Natural history]
I. Title.   II. Series:Bryant-Mole, Karen.   Picture this!
QH48.B89   1999
508--dc21   99-10799
                                                                              CIP

Acknowledgments
The Publishers would like to thank the following for permission to reproduce photographs. Bruce Coleman ; 4 (left),12 (right), 13 (left) and 20 (right) Hans Reinhard, 8 (right) Allan G. Potts, 9 (left), 17 (bottom) George McCarthy, 9 (right) Gordon Langsbury, 12 (left) Mike McKavett, 16 (both) Kim Taylor, 17 (top) John Shaw, 21 (left) Joe McDonald, 21 (right) P. Clement, Positive Images; 10 (both), 11 (both), Tony Stone Images; 4 (right) Pat Bates, 5 (left) G. Ryan & S. Beyer, 5 (right) David Paterson, 8 (left) Tom Ullrich, 13 (right) and 20 (left) Laurie Campbell.

Some words appear in bold, **like this**. You can find out what they mean in the glossary.

Contents

The Country

There are many different kinds of country.

woods

mountains

wetlands

farmland

5

Looking Around

There are many things to see in the country.

You could use these objects to look at and **record** the things you see.

Animals

Here are a few of the many wild animals that live in the country.

fox

mouse

squirrel

deer

Wildflowers

Wildflowers grow without
any help from people.

These wildflowers **bloom** in the spring.

Birds

Different birds like different types of country.

fields

woods

mountains

water

13

Grasses

In our yards, grass is cut short.

In the country, grass is usually left to grow.

14

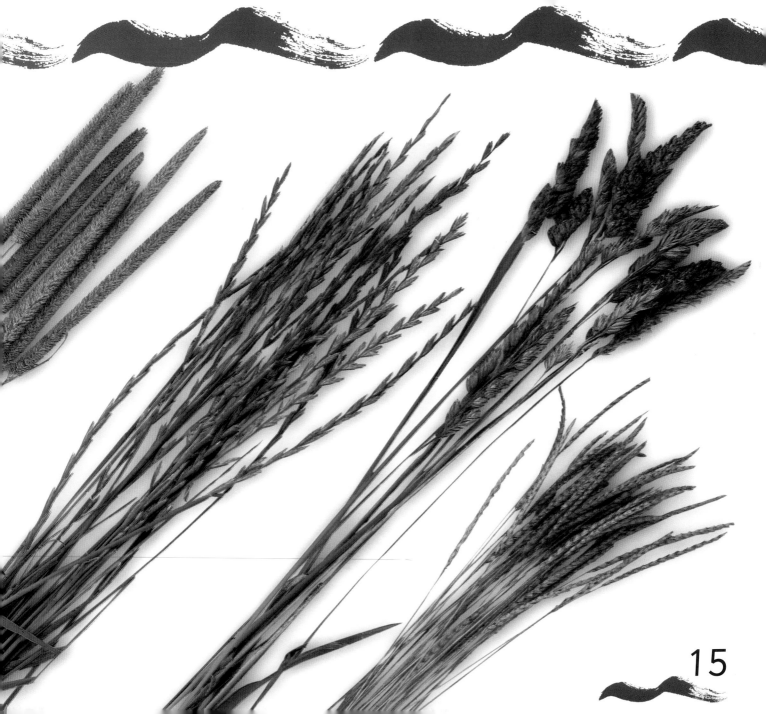

15

Insects

There are many insects in the country.

Have you ever seen any of these insects?

dragonfly

bee

butterfly

grasshopper

Trees

You can **identify** trees by their **bark**, leaves, seeds, berries, **buds**, or flowers.

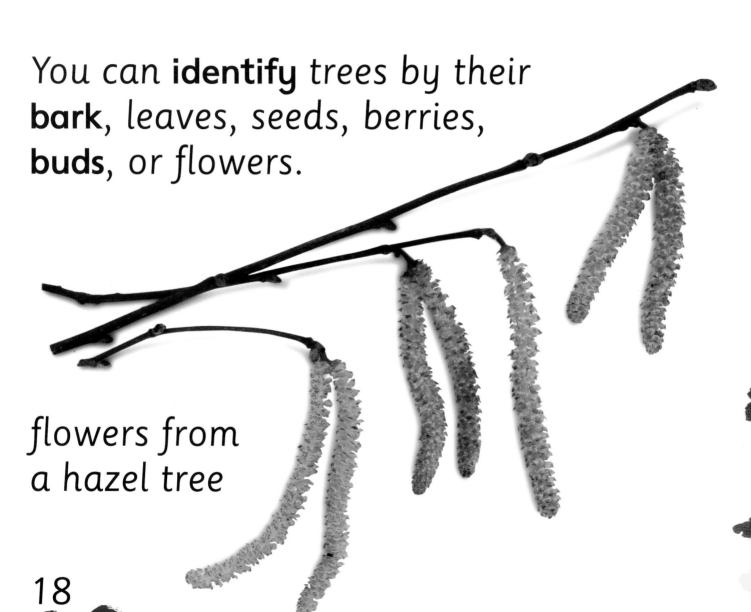

flowers from
a hazel tree

buds from
an ash tree

leaves and berries
from a hawthorn tree

Homes

Wild animals make their homes
in all kinds of places.

under the ground

on branches

in tree trunks

Snails carry their homes with them!

Collections

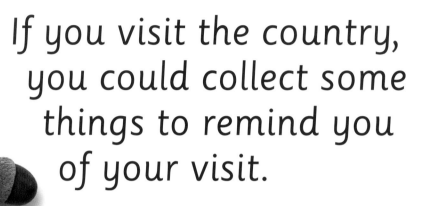

If you visit the country, you could collect some things to remind you of your visit.

Don't bring back
anything that is alive
or pick anything that
is still growing.

Glossary

bark   the outside covering of a tree
bloom   when flowers open up
buds   parts of trees from where new leaves grow
identify   tell what something is
record   write about or draw
wetland   land that is soaked with water

Index

More Books to Read

Chambers, Catherine. *Grasses*. Chatham, NJ: Raintree Steck-Vaughn. 1996.

Stone, Lynn M. *Woodlands*. Vero Beach, FL: Rourke Corporation. 1996.

24